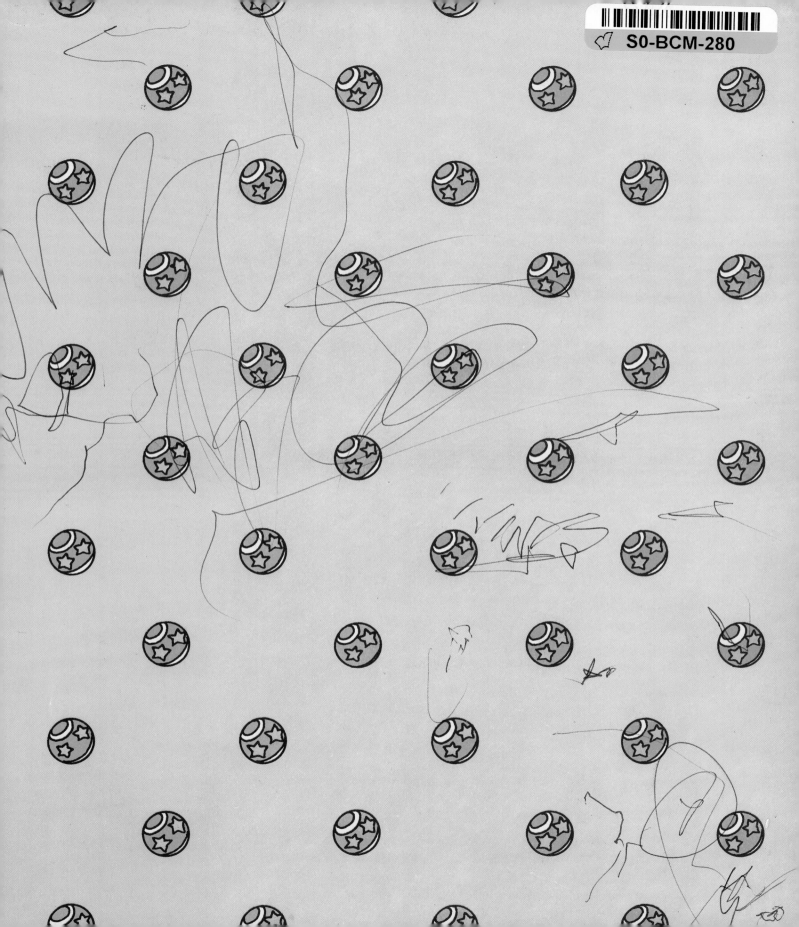

READY, SET, GO!™
A Preschool Language Skills Book
Shaping Up

Written by Ethel Drier   Illustrated by Val Meler

MODERN PUBLISHING
A Division of Unisystems, Inc.
New York, New York 10022

TRIANGLE

A triangle has three sides.

My tent is shaped like a **triangle**.

It's fun to camp out in a tent.

The sail is triangular.

Let's sail our boats!

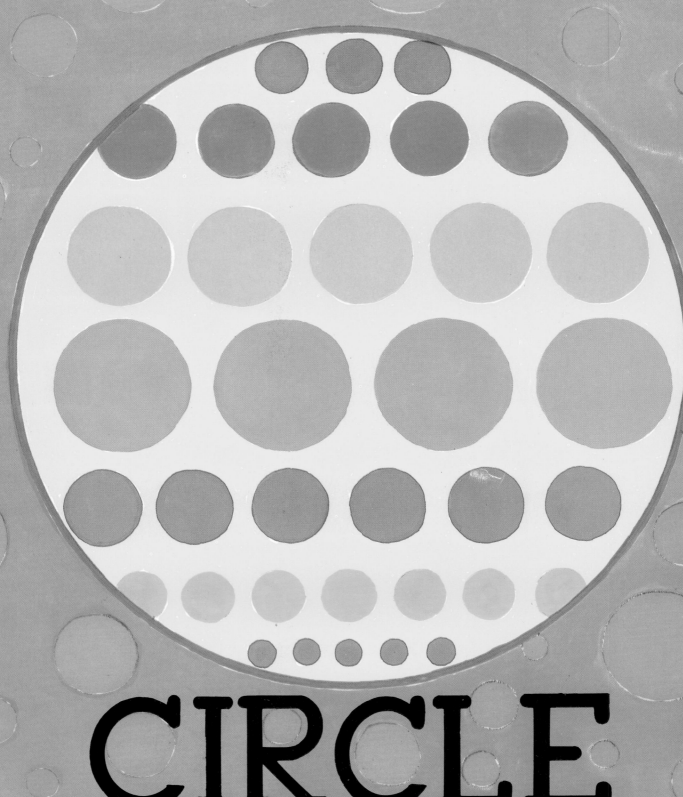

CIRCLE

A circle is round.

The beach ball is shaped like a **circle**.

Let's toss the beach ball.

The drum is circular.

Bang the big drum!

SQUARE

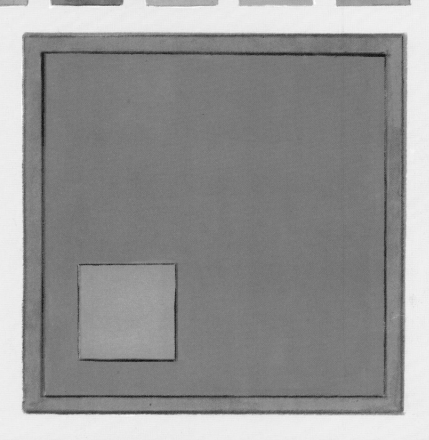

A **square** has four equal sides.

My sandbox is shaped like a square.

We dig in our sandbox.

Square-shaped blocks.

Let's build a tall tower.

DIAMOND

A **diamond** comes to a long point

on the top

and the bottom.

My kite is shaped like a **diamond**.

Our kites fly high in the sky!

These windows are **diamond-shaped**.

Look what's outside our playhouse windows!

RECTANGLE

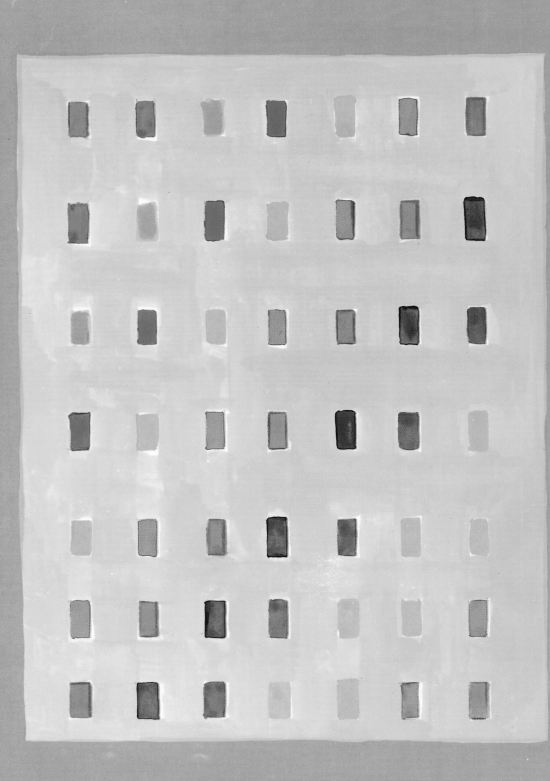

The **rectangle** has two equal, long sides

and two equal, short sides.

My wagon is shaped like a **rectangle**.

It's fun to take rides in a wagon.

This piano is **rectangular**.

Let's play a tune and sing!

The **oval** looks like a long circle.

These balloons are shaped like **ovals**.

Let's have a party
with balloons !

An oval-shaped pond.

Look at the fish in the pond!

Here is a **circle**,

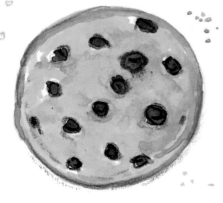

an **oval**,

a **square**,

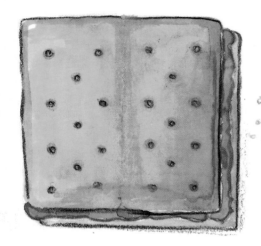

a **rectangle**,

a **triangle**,

and a **diamond**.

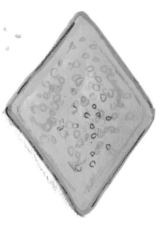

Let's draw all the shapes.

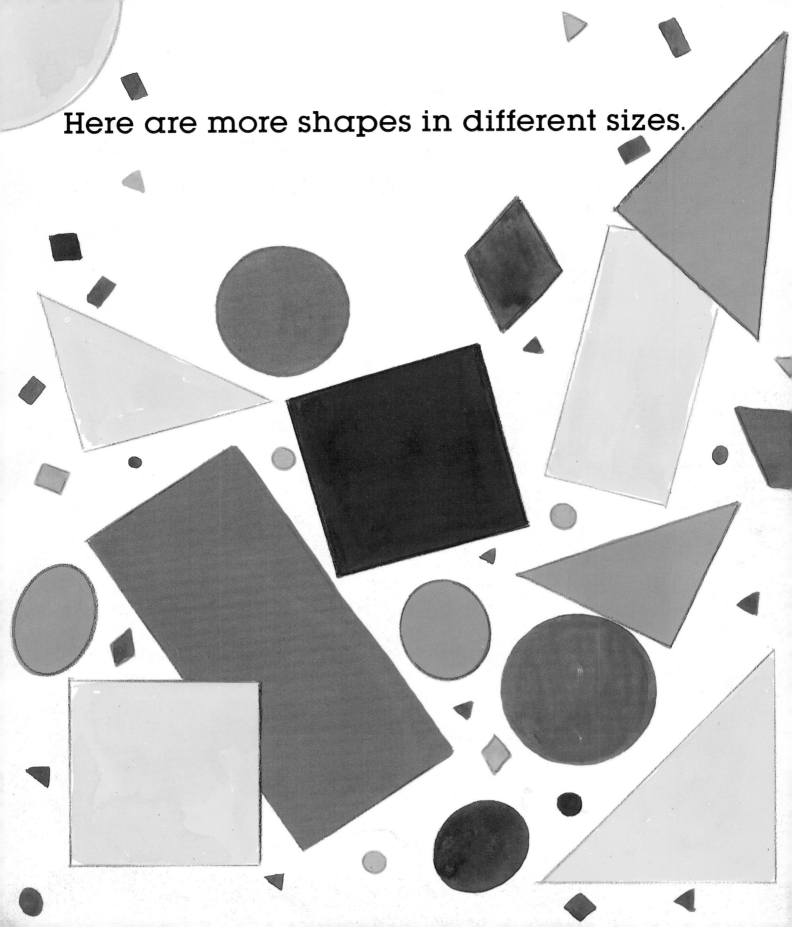

Here are more shapes in different sizes.

Shapes can be
put together to make
other kinds of shapes!